SpringerBriefs in Molecular Science

Chemistry of Foods

Series editor

Salvatore Parisi, Industrial Consultant, Palermo, Italy

The series Springer Briefs in Molecular Science: Chemistry of Foods presents compact topical volumes in the area of food chemistry. The series has a clear focus on the chemistry and chemical aspects of foods, topics such as the physics or biology of foods are not part of its scope. The Briefs volumes in the series aim at presenting chemical background information or an introduction and clear-cut overview on the chemistry related to specific topics in this area. Typical topics thus include: - Compound classes in foods – their chemistry and properties with respect to the foods (e.g. sugars, proteins, fats, minerals, …) - Contaminants and additives in foods – their chemistry and chemical transformations - Chemical analysis and monitoring of foods - Chemical transformations in foods, evolution and alterations of chemicals in foods, interactions between food and its packaging materials, chemical aspects of the food production processes - Chemistry and the food industry – from safety protocols to modern food production The treated subjects will particularly appeal to professionals and researchers concerned with food chemistry. Many volume topics address professionals and current problems in the food industry, but will also be interesting for readers generally concerned with the chemistry of foods. With the unique format and character of Springer Briefs (50 to 125 pages), the volumes are compact and easily digestible. Briefs allow authors to present their ideas and readers to absorb them with minimal time investment. Briefs will be published as part of Springer's eBook collection, with millions of users worldwide. In addition, Briefs will be available for individual print and electronic purchase. Briefs are characterized by fast, global electronic dissemination, standard publishing contracts, easy-to-use manuscript preparation and formatting guidelines, and expedited production schedules. Both solicited and unsolicited manuscripts focusing on food chemistry are considered for publication in this series.

More information about this series at http://www.springer.com/series/11853

Caterina Barone · Marcella Barbera
Michele Barone · Salvatore Parisi
Izabela Steinka

Chemical Profiles of Industrial Cow's Milk Curds

 Springer

Caterina Barone
Associazione "Componiamo il Futuro"
   (COIF) Palermo
Palermo
Italy

Marcella Barbera
DEMETRA Department
University of Palermo
Palermo
Italy

Michele Barone
Associazione "Componiamo il Futuro"
   (COIF) Palermo
Palermo
Italy

Salvatore Parisi
Industrial Consultant, FSPCA PCQI
Palermo
Italy

Izabela Steinka
Gdynia Maritime University
Gdynia
Poland

ISSN 2191-5407   ISSN 2191-5415   (electronic)
SpringerBriefs in Molecular Science
ISSN 2199-689X   ISSN 2199-7209   (electronic)
Chemistry of Foods
ISBN 978-3-319-50940-2   ISBN 978-3-319-50942-6   (eBook)
DOI 10.1007/978-3-319-50942-6

Library of Congress Control Number: 2016960310

Printed on acid-free paper

This Springer imprint is published by Springer Nature
The registered company is Springer International Publishing AG
The registered company address is: Gewerbestrasse 11, 6330 Cham, Switzerland

Contents

Chapter 1
Optimising Lactic Acid Cheese Packaging Systems

Caterina Barone, Marcella Barbera, Michele Barone, Salvatore Parisi and Izabela Steinka

Abstract On the basis of a previous research, it seems that foil-wrapped tray solutions are not particularly chosen by lactic acid cheese consumers. With relation to this study, almost half of the respondent population would have expressed the desire of different packages. Parchment packages and poly(ethylene-co-vinyl acetate)/polyvinylidene chloride/poly(ethylene-co-vinyl acetate) laminates would be removed by 25.0 and 12.5% of customers, respectively. Polyamide/polyethylene double and single packaging would be removed from the market only by 5.0% of respondents. Data have shown that cheese and packaging quality are dependent on lactic acid cheese surface microflora. The type of this microflora is particularly dependent on packaging air-tightness. Anaerobe microorganisms and their metabolites influence properties of packaging materials. On these bases, and considering customers' requirements, a modified packaging system was elaborated. One strategy for optimising traditional packaging systems is the aloe incorporation into the packaging design. The proposed modification of cheese packaging system would include the coating of packaging surfaces (made of synthetic materials) with a thin layer of aloe aerosol and an edible layer, decomposed by lactic fermentation bacteria after a given period of cheese storage.

C. Barone (✉)
Associazione "Componiamo il Futuro" (COIF) Palermo, Palermo, Italy
e-mail: fct1970@libero.it

M. Barbera
DEMETRA Department, University of Palermo, Palermo, Italy
e-mail: marcellapu@hotmail.it

M. Barone
Associazione "Componiamo il Futuro" (COIF) Palermo, Palermo, Italy
e-mail: fct1970@libero.it

S. Parisi
Industrial Consultant, FSPCA PCQI, Palermo, Italy
e-mail: drparisi@inwind.it

I. Steinka
Gdynia Maritime University, Gdynia, Poland
e-mail: ize13ste@wp.pl

© The Author(s) 2017
C. Barone et al., *Chemical Profiles of Industrial Cow's Milk Curds*,
Chemistry of Foods, DOI 10.1007/978-3-319-50942-6_1

1

Keywords Lactic acid cheese · *Lactococcus* spp. · Packaging system · Polyamide · Polyethylene · Polystyrene · *Staphylococcus aureus*

Abbreviations

CFU	Colony forming unit
LAB	Lactic acid bacteria
PA	Polyamide
PE	Polyethylene
PA/PE	Polyamide/polyethylene
PLA	Polylactide
PS	Polystyrene
EVAC/PVDC/EVAC	Poly(ethylene-co-vinyl acetate)/Polyvinylidene chloride/poly(ethylene-co-vinyl acetate)

1.1 Consumer Preferences Concerning Lactic Acid Cheese Packaging

Changes in lactic acid cheese quality during household and non-household storage may have a vital influence on the safety of these products. In detail, the quality of these cheeses may vary depending on the type of packaging used. The level of knowledge of an average consumer does not allow for the proper evaluation of cheese quality. The purchase is made on the basis of packaging appearance (Parisi 2016). The packaging is usually opaque; consequently, an organoleptic evaluation by the consumer is difficult.

Research concerning the determination of the influence of various factors connected to packaging on the choice of lactic acid cheese, and its purchase by consumers were done in 2002 and 2014. The conclusions of this research varied greatly (Steinka and Kukulowicz 2003).

By means of a survey conducted in 2002, we have evaluated the frequency of lactic acid cheese purchase, the influence of the packaging qualities, information contained therein and the frequency of this cheese purchase based on type of packaging (Steinka and Kukulowicz 2003).

Respondents were women who ran or participated in running a household. These women had various degrees of education (age range: 18–60 years).

Research results indicated that over half the respondents (61.4%) purchased lactic acid cheese two or more times a week. Respondents evaluating the influence of packaging on the purchase of products constituted 86.1% of all respondents, including respondents who purchased it once a week (Steinka and Kukulowicz 2003).

The evaluation of preferences concerning qualities and characteristics of packaging materials has shown that the most important factor influencing the choices made by customers was the tightness of packaging. This quality was mentioned by as many as 71.8% of respondents. Over 59.4% of respondents stated that among factors deciding on the purchase of lactic acid cheese, a large role is played by the aesthetics of the packaging. For 46.6% of respondents, the information about the manufacturer identity and product brand was very important. Among the factors influencing the choice of lactic acid cheeses, visual stimuli connected with the quality of packaging played a role with over 59.0% of respondents.

Nutritional factors connected to the informative function of the packaging were at the fourth place among parameters influencing the choice of cheeses. The easy package opening was a factor that influenced the choice and purchase of cheeses in almost 30.0% of respondents.

Among the 14 listed factors with possible influence on choice of lactic acid cheese, aside from the qualities of packaging itself, over 58.4% of respondents listed the product price (Steinka and Kukulowicz 2003).

In addition, the conducted research showed that package types preferred by the customers were correspondent to a polystyrene tray wrapped with polyethylene foil and vellum. The choice of this type of packaging indicates a lack of connection between preferred qualities of packaging (tightness) and consumer habits. Choices respond to the consumer need of packaging that is easy to open.

The survey had shown, among the packaging qualities, only two out of 14 factors that concerned informative function had any influence on the decision of product purchase. Obtained data also showed that the biggest role in product purchase is played by functional and visual qualities of the packaging, and as for qualities not connected to type of packaging—the decisive factor is price (Parisi 2016; Steinka and Kukulowicz 2003).

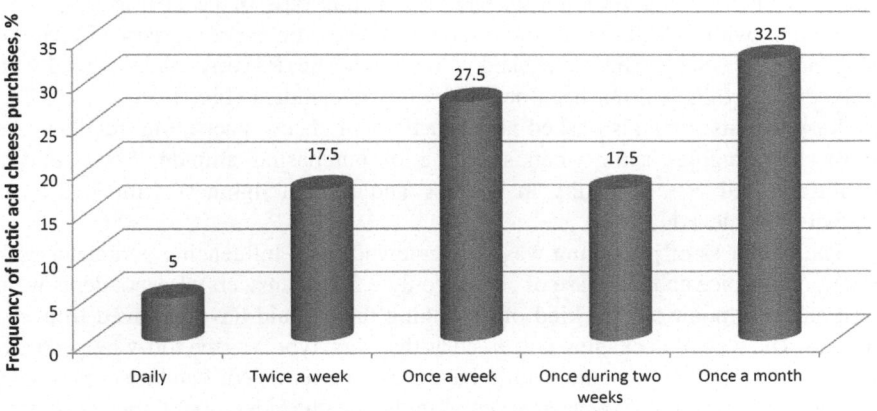

Frequency of lactic acid cheese purchases

Fig. 1.1 Results of a research on consumers' attitude with relation to lactic acid cheeses. Frequency of related purchases is displayed versus time intervals between subsequent purchases

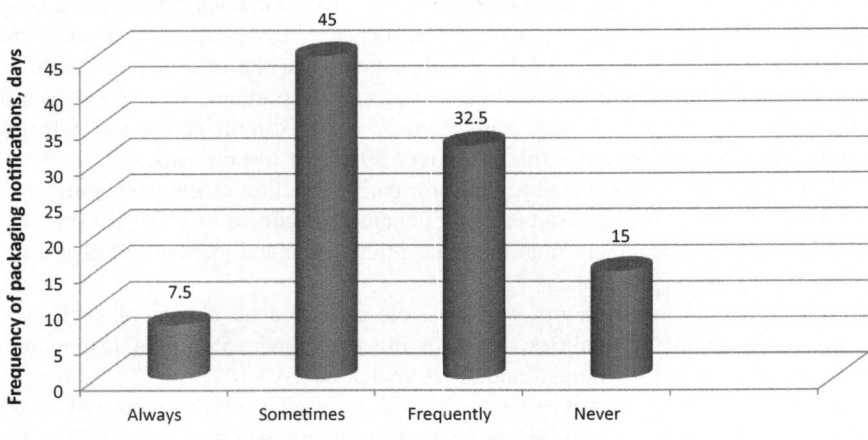

Fig. 1.2 Results of a research on consumers' attitude with relation to lactic acid cheeses. The influence of packaging materials on consumers' purchases is measured by means of the frequency of packaging observations

A subsequent research conducted in 2014 on 40 random respondents has shown that infrequent buyers were dominant: once a week and once a month (Fig. 1.1). According to these data, 32.5% respondents noticed cheese packaging often; in addition, 7.5% of respondents always noticed the type of packaging (Fig. 1.2). Results of this survey allowed determining that the packaging system preferred by the customers was a double-packaging solution. In this system, cheeses are wrapped in parchment, and then hermetically packed with polyamide/polyethylene (PA/PE) films. Over 50.0% of respondents had chosen this type of packaging (Fig. 1.3). Packaging with PA/PE films only ranked second favourite system (27.5% of respondents). Cottage cheeses packed with polystyrene (PS) trays and wrapped with polyethylene foils had not been named by any respondent (Fig. 1.3).

Respondents were also asked about lactic acid cheese packaging qualities, the most discouraging factor when speaking of purchasing attitude. Some of the qualities listed were difficulty in opening and lack of tightness with 72.5%, as shown in Table 1.1.

The thickness of packaging was not perceived as an influencing parameter with relation to choice and purchase of cheese (only 2.2% of answers). Respondents were also asked to point out the kind of packaging they would have removed from the market. The type of packaging perceived as the worst type was the foil-wrapped tray. Almost 48.0% of respondents would have that type removed. Parchment packages and poly(ethylene-co-vinyl acetate)/polyvinylidene chloride/poly(ethylene-co-vinyl acetate) (EVAC/PVDC/EVAC) laminates would be removed by 25.0 and 12.5% of customers, respectively. On the other side, PA/PE double and single packaging would be removed from the market by 5.0% of respondents (Table 1.2).

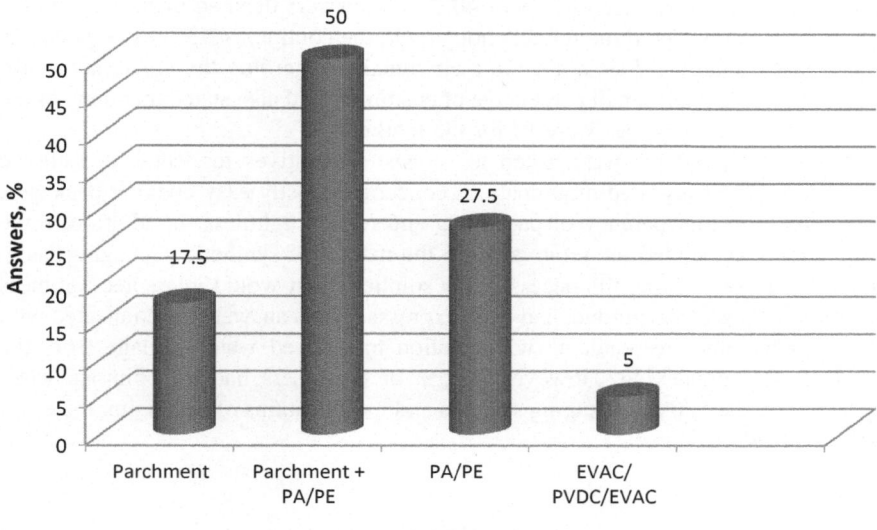

Fig. 1.3 Results of a research on consumers' attitude with relation to lactic acid cheeses. The influence of packaging materials on consumers' purchases is measured by means of the frequency of packaging observations

Table 1.1 Lactic acid cheese packaging properties that are the most discouraging when it comes to purchase

	Lactic acid cheese perceived properties						
	C	S	T_h	T_r	T_g	L	D
Percentage of respondents (%)	5	10.0	2.2	5.0	7.5	25.0	47.5

C colour, S smell, T_h thickness, T_r transparency, T_g tightness and L lack of tightness; D difficulty in opening

Table 1.2 Packaging types. Perceived preference amounts by respondents

Packaging types for lactic acid cheeses. Preference amounts					
	Parchment	Parchment +PA/PE foil	PS tray wrapped with PE foil	EVAC/PVDC/EVAC laminate	PA/PE laminate
Percentage of respondents (%)	25.0	10.0	47.5	12.5	5.0

Respondents were asked for a subjective evaluation of adequacy of packaging choice for lactic acid cheeses. About 40.0% of answers deemed them appropriate, while 22.5% considered them well enough. On the contrary, 42.5% of respondents stated that packaging does not serve their function regarding the quantitative protection of products. A small percentage of customers had also stated that none of the proposed packaging types were fit for the market.

Finally, respondents were asked to suggest alternatives for lactic acid cheese packaging. The suggested modifications concerned mostly easy opening packages, the increase in transparency of packaging and related tightness. In addition, some respondents demanded the removal from the market of synthetic packaging materials. At the same time, the necessity for solutions that would guarantee tightness after unpacking of the product had been expressed. This answer was connected with demands by other respondents with relation to reduced water spillage from the product after unpacking. However, 82.5% of customers had no opinions when speaking of alternative packaging systems and modifications of packaging materials or assembled packages.

1.2   Characteristics of Packaging Used for Lactic Acid Cheese and Cottage Cheese

The type of material used in packaging has a large influence on longevity and safety of cheeses. There are several types of packaging materials applied to this type of product. The most traditional material used is parchment. Some dairy manufacturers have recently started to use also parchment metallised with aluminium. The lack of tightness for this type of material constitutes its biggest drawback when speaking of cheeses.

For cottage cheeses, manufactured with lactic acid bacteria (LAB) and rennin (as opposed to lactic acid cheese made using only LAB), the type of packaging often used is made of polystyrene (Fig. 1.4). At present, a double-packaging system

Fig. 1.4 Polystyrene (PS) packaging for cottage cheese

Fig. 1.5 A cheese packaged
with EVAC/PVDC/EVAC
laminates

using various materials is also considered: parchment is the closest layer to cheese surfaces, while the outside layer is made of synthetic materials (PA/PE, sometimes EVAC/PVDC/EVAC laminates).

For cheese packaged in non-vacuum systems, the most common materials used are non-transparent, white EVAC/PVDC/EVAC laminates with transverse bindings, manufactured using the co-extrusion method. This material is heat-shrinkable (Fig. 1.5). In addition, PA/PE packaging films with various thickness and barrier characteristics (gases and water vapour) are also available on the market. With relation to PA/PE laminates, the most common choice has generally thickness values around 80 µm (dimensions: 15 cm × 25 cm, Fig. 1.6). Barrier features show that optimum foils are those with vapour permeability equal to 1.7 g/m²/d, while oxygen and carbon dioxide permeability parameters are 56 and 168 cm³/ m² × 24 h × bar, respectively, for carbon dioxide; all data are measured at 23 °C, 75% relative humidity (Steinka 2005a).

Fig. 1.6 A cheese packaged
with PA/PE materials

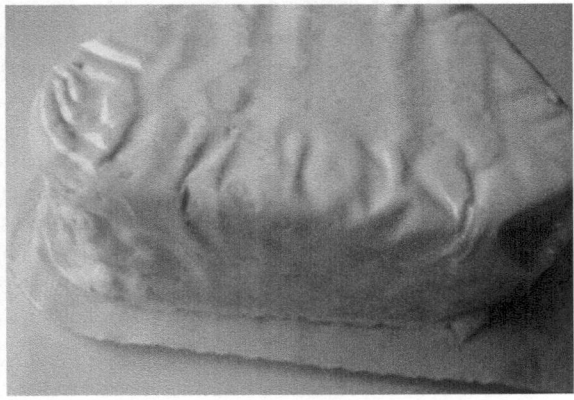

Also polystyrene trays are used for cheese packaging. However, trays with packaged cheese should not be turned, due to the possibility of crushing products. The ample space between the layer used for wrapping and the product itself, contains a sizable concentration of oxygen. Only cheeses with low syneresis levels may be packaged in this way. Expiration date for cheeses packed on trays is much shorter in comparison with other packing methods.

1.3   Influence of Physical Qualities of Packaging Materials on Packaging Functionality

According to above-mentioned survey results (2002 and 2014), tightness appeared to be the most important packaging factor for customers. In particular, packaging tightness was as important for customers as the informative function of the packaging. Tightness, which realises consumer requirements regarding the easy transportation of packaged cheeses without water spillage, is one of the two postulated qualities named in first place by respondents.

Moreover, lactic acid cheese may influence packaging material qualities during storage periods. Actually, this reflection concerns inappropriately chosen package materials, with regard to product type and packaging system. With relation to products, not only the content of fats and acidity plays a part, but also the clot structure, connected to water amount.

An important quality of lactic acid cheeses is the fluctuation of the water phase contained in the whey that may flow outside of the product. Different whey quantities, coming in contact with the packaging surface, are responsible for changes in packaging material qualities during a prolonged storage period. This is an important issue during a prolonged storage period of cheeses. In addition, changes in packaging barrier qualities (these modifications may occur after 21 days of storage) are dependable on the type of packaging material.

Water circulation between cheese and packaging surfaces is also dependable on barrier qualities of packaging materials (Steinka and Parisi 2006), with important consequences when speaking of microflora growth on superficial layers. In detail, water which is close between cheese surface and package can determine detectable modifications of barrier properties.

At the same time, the metabolism of technological microflora (*Lactoccocus* spp. and *Leuconostoc*) and microorganisms present on cheese surfaces also contributes to changes, not only with relation to barrier properties. We have shown barrier properties of various materials, evaluated as potential lactic acid cheese packaging, in Table 1.3. Swelling tests on packaging materials are one of the measures of water circulation between cheese and packaging surfaces (Fig. 1.7). Moreover, water vapour may be cause of packaging swelling on superficial product areas.

	Barrier properties for packaging materials, expressed as % permeability variation. Control after ...		
Packaging type	0 days	7 days	14 days
PA/PE	0.0	0.042	0.11
EVAC/PVDC/EVAC laminate	0.0	0.039	0.09
Parchment	0.0	0.0	0.0
Metallised parchment	0.0	0.0	0.0
PS	0.0	0.0	0.0

Table 1.3 Permeability changes of packages used with lactic acid cheese versus water vapour

Experimental data (Steinka, unpublished research)

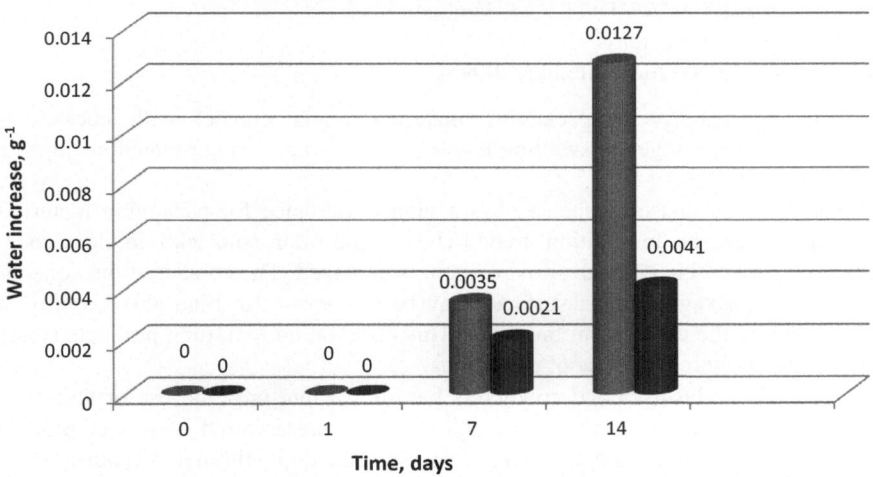

Modification in packaging materials:
EVAC/PVDC/EVAC laminate and PA/PE materials

Fig. 1.7 Modification in packaging laminates after prolonged contact with lactic acid cheese

The effect of water absorption by the packaging may also be correlated with changes in the mass of packaging laminates. Tests have shown this result after 14 days for EVAC/PVDC/EVAC laminates and after 7 days for PA/PE packaging materials (Fig. 1.7).

In general, the most commonly used packaging system for lactic acid cheeses is vacuum packaging. Data shows that the use of lowered pressure (values around 23 kPa) for cheese packaging may lead to decrease in water loss by-products (Mazur et al. 2011). Mazur and coworkers have shown that only 3% of water loss is observed after 20 days of storage with similar pressure values for cheese packaging (Mazur et al. 2011). The value of parameters of pressure applied in vacuum packaging determines the content of water in stored cheeses (Mazur et al. 2012). In this situation, such small water loss may make for a significant stability of barrier qualities of packaging materials.

A comparison of barrier properties changes of various packaging in interaction with metabolites of the cheese microflora is shown in Table 1.3.

In addition, the quality of products and the stability of container features are largely dependable on fluctuations in barrier properties of packaging materials: these variations are influenced by stored cheeses. Changes of these properties during storage time can be described by means of different equations such as

$$M_c = 0.0028x^2 - 0.005x + 0.0022\left(r^2 - 1\right) \tag{1.1}$$

$$M_p = 0.0021x - 0.002\left(r^2 - 0.998\right) \tag{1.2}$$

where

- M_c is for: EVAC/PVDC/EVAC laminate mass
- M_p is for: PA/PE mass
- x is for: storage time (Steinka 2009a).

The absorbability of packaging materials under contact with cheeses is responsible for changes in swelling levels of packaging, and correlated changes in mass (Figs. 1.4 and 1.5).

The aqueous phase of cheeses has a vital importance for packaging materials such as parchment. In addition, traded cheeses are often sold with an extra packaging for preventing damages if parchment is damaged. The contact of the aqueous phase with packaging paper surfaces may be the reason for high absorbability of this layer and the consequent breakage. This observation regarding products stored for 24 h after purchase in cool conditions.

Research by Jasinska and coworkers has shown that only the use of laminates with thickness of 100 microns can allow for the preservation of proper product structure and texture, when speaking of cheese packed in vellum and synthetic foils (Jasinska et al. 2014). These data have confirmed our previous observations: storage of cottage cheese packed in synthetic foils should not last longer than 7 days. However, a 3 weeks period of expiration is applied by manufacturers in practice (Jasinska et al. 2014; Steinka 2007a; Steinka and Kukulowicz 2004a).

Fulfilling of consumer expectations regarding ease of packaging opening is a tough issue. Lactic acid cheese characteristics enforce the use of packaging with high barrier qualities, but the ease of opening of the packaging demands materials with low resistance to breakage. The decrease in mechanical properties such as resistance to breakage may imply changes in the resistance of packaging materials. In the case of synthetic packaging, the resistance to breakage is vitally connected with the stability of structure (the stability of crystalline phase).

With reference to PA/PE materials, tests performed of the inside laminate layer in contact with the water phase of lactic acid cheeses have exhibited changes in the structure. Observations using a Nikon Alphaphot 2 YS2 polarising microscope in

polarised reflected light (for surfaces) and penetrating polarised light (for observation of cross-sections) allowed determining slight changes in the structure of this
packaging material (Steinka and Kukulowicz 2004a).

Research performed with penetrating polarised light allowed for observation of
modifications in the crystalline phase of polyamide (PA) and polyethylene
(PE) polymers. However, in contract to PE, which could be deemed a
semi-crystalline polymer characterised by small crystalline phase content, PA
reacted to contact with lactic acid contained within the cheese aqueous phase
(Steinka and Kukulowicz 2004a). Lactic acid (solution pH: 4.7) caused the variation of penetrating light beam size within the duration of storage at 4 °C
(Table 1.4). It has been determined that the influence of the water phase of cheeses
may cause changes in the crystalline phase of used laminates (Figs. 1.8, 1.9 and
1.10).

The high degree of correspondence between the time of influence of lactic acid
(pH: 4.7) and the modification in the structure of packaging materials has been
verified by means of a linear correlation (Eq. 1.3). Changes in the penetrating light
beam determined the behaviour of the crystalline phase of the copolymer (Steinka
and Kukulowicz 2004a). Correlation has been determined between the length of
light beam and contact times of the laminate with lactic acid:

$$Y = -70.3x + 240.5 \, r^2 - 0.991 \qquad (1.3)$$

Table 1.4 Changes in beam diameter of polarised light during interaction of lactic acid cheese—
PA/PE material

Time of packaging interaction with lactic acid, days	Diameter of penetrating light beam, μm
1	173.9
4	92.5
14	33.3

Experimental data (Steinka, unpublished research)

Fig. 1.8 PA/PE laminates
after the first day of contact
with lactic acid

Fig. 1.9 PA/PE laminates
after 7 days of contact with
lactic acid

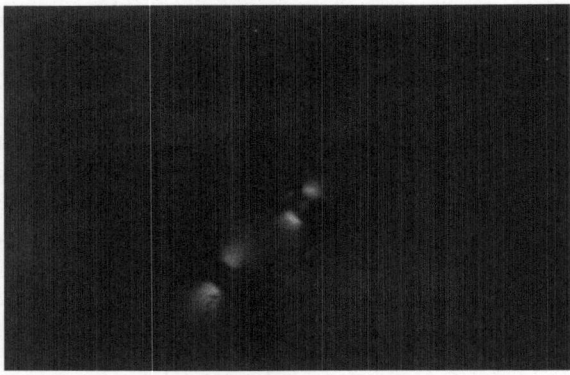

Fig. 1.10 PA/PE laminates
after 14 days of contact with
lactic acid

where:

- Y is for: diameter of light beam
- x is for: time of contact of laminate with lactic acid.

1.4 Microbiological Interactions Versus Quality of Packaging Used for Lactic Acid Cheese

Tests on synthetic packaging materials after contact with stored cheeses have confirmed that interactions between packaging surfaces and products have a vital significance for both cheeses and their packaging. During storage of products with high water content, adherence of cheese superficial microflora to the inside surface of packaging walls might take place. The adherence of microorganisms to surface is dependent on the said surface, on environmental conditions and on the hydrophobic nature of inner structures within the cellular wall of life forms. The superficial microflora isolated from cheeses might be ascribed to hyphae type fungi (moulds), yeast, enterococci, coliform bacteria, and staphylococci. Such impurities within the cheese product are variegated depending on the type of used packaging and its barrier qualities, according to our previous research (Steinka 2005b, 2006, 2007a, b, 2008a, 2009b, 2012; Steinka and Blokus-Roszkowska 2009; Steinka and Morawska 2007; Steinka et al. 2011).

The presence of microflora on the inside wall of packaging may provide an additional determining factor for changes in stability of packaging materials. Modifications in the location of product fragments during any operations connected with transport cause also the shift of superficial microflora. Adherence of microorganisms to packaging surfaces may also be significant for the quality of stored products and containers. Observed phenomena are probably connected to the change in polarity of surface layers of interested life forms. They are stimulated by modifications in acidity and water activity of cheese surfaces during product storage.

Our research has proven that there is no clear answer to the question about the adhesive capabilities of microorganisms on packaging surfaces used for lactic acid cheese (Steinka 2008b). A peculiar importance should be ascribed to factors such as fat and water content in packaged products. In particular, there was observed an ability to form biofilms by *Lactococcus* spp. and *Enterococcus* spp. when speaking of materials with average fat contents and high water amounts. At the same time, yeasts did not form biofilms under these conditions. A negligible growth on laminate surface was observed for moulds (Steinka 2008b).

However, the level of adhesion of microflora to packaging is more variegated in case of multi-species populations and increased fat contents in cheeses than for leaven microorganisms (Steinka 2005c; Steinka and Kukulowicz 2004b). The adherence of streptococci of the *Lactococcus* spp. type to PA/PE surfaces during cheese storage may be described with a third-degree polynomial equation. This behaviour is a proof of complicated processes taking place between microorganisms and packaging laminate superficial areas.

Adherence of microflora to container surfaces may cause mechanical changes in packaging features (Steinka 2007c, 2012). This situation pertains to both techno-logical microflora and microorganisms constituting microbiological impurities of products. Long-term presence of microflora on the surface of packaging may be, besides the fluctuation of the water phase, a determinant of changes in packaging resistance and mass (Steinka and Kukulowicz 2004a).

The form of equation necessary to assess number and type of responsible microorganisms for physical changes of chosen packaging materials is presented below (Mazur et al. 2012; Morawska 2012):

$$\log N_p = \log[(M_{PA/PE}) \times k] \tag{1.4}$$

$$\log N_p = \log[(M_{EVAC/PVDC/EVAC}) \times k] \tag{1.5}$$

where

- N_p is for: number of microorganisms in lactic acid cheese packed with vacuum systems
- N_b is for: number of microorganisms in lactic acid cheese packed with EVAC/PVDC/EVAC laminates
- k is for: a coefficient equal to 10^4 (Steinka 2009a).

Our earlier research on product/packaging interactions have shown that the activity of *Staphylococcus aureus* and coagulase-negative staphylococci have a significant influence on changes in EVAC/PVDC/EVAC laminate mass after the storage process of cheese (Steinka 2007b). Changes in PA/PE resistance to breakage were also observed during the contact of this packaging with *Lactococcus* spp. type bacteria, with inoculum stimulating the conditions within cheeses. The evaluation of package stability has shown, among other things, that biodegradable materials such as polylactide (PLA) do not contribute to longevity of cheese storage in cool conditions (Morawska 2012).

Packaging stability is dependent on the degree of coverage of packaging materials by microorganisms. With reference to biodegradable packaging materials such as PLA, the degree of coverage was the biggest value (Table 1.5). This result is an important reason for which prolonged storage of products in PLA packaging is not recommended.

Research by Morawska has confirmed that biogenic substances constituting the products of metabolism of technological microflora of cheese are responsible for triggering changes in traditional cheese packaging such as PA/PE (Morawska 2012). Research performed in a model system allowed to conclude that modifications in parameters describing packaging quality have to be ascribed to the aqueous phase, microorganisms, and related metabolic products. Our previous research

Table 1.5 Degree of coverage of packaging material surface by microorganisms

Packaging type	Degree of surface coverage after interaction [%]	Microorganism
PA/PE	56	*Staphylococcus aureus*
EVAC/PVDC/EVAC laminate	2	
Polylactide(PLA)	72	
Parchment	Dispersed agglomerations of colonies	
PA/PE	48	*Lactococcus spp.*
EVAC/PVDC/EVAC laminate	13	
PLA	87	
Parchment	Uneven coverage dependent on the size of aqueous phase	

Experimental data (Steinka, unpublished results, year 2013)

indicated that water phase in cheeses influences notably properties of packaging materials (Steinka and Morawska 2006; Steinka and Parisi 2006). In addition, properties such as breakage tension changed during the time of storage of cheese clots and whey, both in traditional packaging materials (PA/PE laminates) as well as the proposed PLA (Morawska 2012).

Research by Morawska on qualities of possible packaging materials such as PLA has shown that polylactide may be competitive if compared with traditional synthetic packaging materials. Literature shows that the hydrolysis of this material happens under the influence of high hydrogen ion concentrations or in presence of lactic acid (various forms). Morawska's research has additionally shown that PLA foil exhibits decrease in resistance to tension under lactic acid influence (Morawska 2012). This factor disqualifies the material in terms of packing cheese intended for long-time storage.

1.5   Optimization of Packaging Used for Lactic Acid Cheeses. Suggestions

Usage of traditional non-hermetic packaging materials does not contribute to acceptance by most part of consumer respondents. This situation is connected to the observed lack of consumer comfort during cheese transportation. Non-hermetic packaging in parchment causes larger leaks of whey outside the container. Modification by using double-packaging causes possible leakages of the water phase between packaging layers (Fig. 1.11).

Ecologic requirements force packaging manufacturers to search for biodegradable packaging materials, but specific features of lactic acid cheese—whey and food acidity—do not help to find a proper solution for this issue. Water bioavailability in the space between two packaging layers may be favourable to microflora shifts within the aqueous medium. Changes in packaging placement during transport may favour microbial penetration into package surfaces. This phenomenon may be observed taking into account storage times as recommended by packaging manufacturers.

Fig. 1.11 Leakage of whey from inner packaging (parchment) to the area between packaging materials

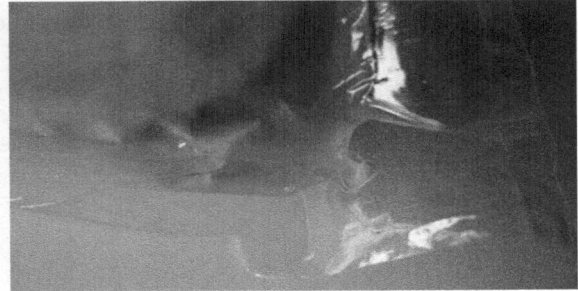

Table 1.6 Changes in *S. aureus* counts during 14 days storage of lactic acid cheese in PA/PE

Plant identification	Stored lactic acid cheese in PA/PE packaging: staphylococci counts, $\log_{10}$CFU/g	
	Initial count (0 days)	Maximum count after 14 days of storage
Plant M	2.67	2.85
Plant K	1.69	2.50
Plant C	1.60	2.47

Experimental data (Steinka's unpublished research, years 2008–2010)

Literature indicates that there are biodegradable materials, such as PLA, which are successfully used for packaging rennet maturing cheeses. However, this type of cheese does not exhibit whey flow. For this reason also, PLA cannot be used for lactic acid cheese packaging.

From the comparison of stability of PLA packaging with those containers made of different synthetic materials, it appears PLA is not suitable for packaging lactic acid cheese, taking into account also microbiological interactions (Morawska 2012).

1.6 Verification of Necessity for Optimization of Packaging Materials for Lactic Acid Cheese

Use of improperly chosen packaging materials made of synthetic materials may imply microbiological changes and numerical increase, when speaking of relatively anaerobes microflora. Their growth has been demonstrated by multiple researches on lactic acid cheese over the last years (Steinka and Kukulowicz 2004a; Steinka et al. 2011).

Growth inhibition in relatively anaerobe bacteria on cheese surfaces is vital, due to the consumer safety. Inside the niche formed between packaging surfaces and the product, there are favourable conditions to the growth of such microorganisms (hermetic containers). This situation pertains especially to *S. aureus*-type bacteria in various layers of cheeses, favoured by the atmosphere present inside containers (Steinka 2006).

Research made in 2008–2010 has shown the presence of staphylococci in hermetically packaged cheeses at the level of 1.60–2.67 $\log_{10}$ Colony Forming Unit (CFU)/g. This level was observed immediately after packaging. In detail, staphylococci population has exhibited variegated behaviour in three researched facilities (Table 1.6). No growth of bacteria has been observed in lactic acid cheese from facility M, while a decrease in number of *S. aureus* has been observed in facility K, by the average of 1.26 $\log_{10}$ CFU/g. Finally, an increase in growth of staphylococci was observed with facility C. These data confirmed that the behaviour of the *S. aureus* population in products with similar counts may show variations immediately after packaging. It should be noted that cheeses were packaged with the same type of material, and the initial number of bacteria was relatively similar. Maximum growth of bacteria was observed after 2 or 4 storage days. It should be assumed that the packaging system requires modifications, taking into account the nesting-type dispersion of those bacteria inside the product (Steinka 2006).

Hermetic packaging also encourages yeast growth, which may stimulate the production of staphylococcal enterotoxins in the context of staphylococci interactions with those fungi. A computer programme simulating those relations has been designed and presented in our earlier works (Steinka 2011).

Research concerning optimisation of packaging systems is available in literature (Steinka 2003; Panfil-Kuncewicz et al. 2006; Steinka and Blokus-Roszkowska 2009). In brief, this activity has concerned the usage of substances with biostatic character or oxygen absorbers. However, use of active packaging materials (oxygen absorbers) is not the best solution because of the interference of the chemical absorbent inside cheese packages. In addition, this solution may be an aesthetic barrier from the consumer's viewpoint. For the manufacturer, this option may be an alternative: it guarantees the possibility of prolonging product durability.

Obtained data shows that the development of oxygen microorganisms may be retarded with use of oxygen absorbents (Panfil-Kuncewicz et al. 2006). However, reduction of oxygen levels below 0.5% obtained by this method may stimulate the growth of relatively anaerobe and strict anaerobe bacteria. This situation pertains to such microorganisms as staphylococci or *Listeria monocytogenes,* the growth of which is not retarded by microaerophilic conditions within the packaging. On the other hand, coliform bacteria do not find lactic acid cheese particularly conducive to growth; consequently, the use of active packaging in order to reduce them does not appear necessary.

Later research on active packaging for lactic acid cheese has shown that their use does not cause any significant reduction of undesired microflora in products (Steinka and Stankiewicz 2002). Fitzgerald and coworkers have shown that qualitative changes in animal by-products are dependent on packaging materials (Fitzgerald et al. 2001). Conclusions of this research lead towards determining that the oxygen profile obtainable for lactic acid cheese with oxygen absorbents, as proposed by Panfil-Kuncewicz and coworkers, may stimulate the growth of anaerobes (Panfil-Kuncewicz et al. 2006). Thus, the safety of lactic acid cheese packed with oxygen absorbents may be under threat.

The search for other packaging types, or the modification of presently existing packaging systems, may be much better than the usage of chemical absorbents. An alternative solution may also be the use of nanoparticles in cheese packaging technology. Other alternatives are coated packaging materials with biostatic substances of the chemical and structural modification of packaging materials.

1.7 Changing the Packaging System. Proposals

The basis for the proposals for changes in the cheese packaging systems is the usage of biostatic substances-coated PA/PE packaging. Our earlier research, aimed at the evaluation of interactions between packaging and cheese surface coated with a layer of aloe aerosol (named as Variant II), has shown that it is an effective safekeeping method. Aloe was prepared in accordance with patent number 19,769 (Steinka 2007b) and the cheese surface was covered, before packaging, with a layer of sterile aloe aerosol (amount: 1 cm^3 aerosol per 1 cm^2 superficial area). After 15 min, products were packed with a vacuum packaging machine using PA/PE bags, and stored in a freezing chamber for 2, 4, 7, and 14 days at temperatures of 4 °C (Steinka 2008c).

Obtained results have shown large effectiveness of aloe aerosol. The effectiveness was observed for all microorganisms. Up to 62% of variation in microbial count ascribed to *S. aureus* has been noted using this method. The effectiveness of activity has been noted regarding staphylococci. A relatively low impact was noted for moulds (Steinka 2003). At the same time, observations were performed concerning changes in population of bacteria responsible for shaping product qualities. The behaviour of bacteria of *Lactococcus* spp.-type bacteria under the influence of applied aerosol could be expressed by means of a polynomial equation:

$$L_a = -1.248 \times 10^6 + 1.008 \times 10^6 \times t + 1.194 \times -64381.85 \times t^2 - 0.043 \times t \times L - 3.768 \times 10^{-9} \times L^2$$

$$(1.6)$$

where

- L_a is for: number of *Lactococcus* spp. in cheeses with aloe addition
- L is for: the number of *Lactococcus* spp. in control group cheese
- t is for: time of cheese storage.

The negative value of the first factor in the above mentioned equation pointed to the direction of changes in the microbial count of *Lactococcus* spp. influenced by aloe, suggesting a reduction in their numbers. This direction of changes would be also desirable from the point of view of retardation of lactic acid cheese overacidity.

The performed research has shown that an addition of aloe to cheeses with variable microbial counts guarantees an improvement in product quality. A reduction in staphylococci counts was observed during the 14 days—period of storage, and also retardation in the growth of filamentous fungi in those cheeses. Results led to conclude that aloe aerosol cannot be used for optimisation of quality of cheeses with high degree of impurities of enterococcus and yeast (Steinka 2003, 2008c).

Assuming the improvement in the performance of the HACCP system within the cheesemaking process, it has to be considered that the level of those latter microorganisms will be low, or they will be absent completely. However, our multiple research regarding *S. aureus* levels in cheeses has shown their presence at an average level of 2 $\log_{10}$ CFU/g. This result entitles the conclusion that the use of aloe aerosol may be a favourable factor for reduction of those microorganisms in cheeses (Steinka 2008c). Also, the research allowed determining that the effectiveness of biostatic activities correlated to sprayed cheese surfaces may be a base for further investigation in this direction.

The proposed modification of cheese packaging systems using this method would include the covering of packaging surfaces made of synthetic materials with a thin layer of aloe aerosol and an edible layer, decomposed by lactic fermentation bacteria after a given period of cheese storage. The proposed algorithm for such a modification would be the following:

- Aerosol preparation according to Patents
- Aerosol spraying on packaging inner surfaces
- Coating of cheese surfaces with an edible layer
- Drying of cheese surfaces
- Final packaging with PA/PE foil.

With relation to this variant (variant I), the edible layer may consist of a casein preparation. In order to obtain biostatic effects in relation to bacteria responsible for food hygiene, a modification needs to be also implemented in the technological line. This modification may pertain to both technological lines which synthetic packaging materials are used in, as well as vellum, in a double-layer packaging system.

The result of technological modifications of packaging was the effect of retardation of unwelcome microflora growth (relatively anaerobe life forms), as shown in Table 1.7.

Results of mathematical modelling, based on empirical data, allowed obtaining interesting conclusions. It has been determined that second-degree polynomial equations describing the behaviour of staphylococci in stored cheeses showed differences when cheeses were sprayed with plant aerosols (variant II).

Due to automation of the performed process (cheese manufacturing), the spraying step on packaging foils with aerosol may be difficult. For this reason, it has also been suggested to cover the cheese surfaces right after pressure process (variant III). In this modification, packaging should be classified as a two-layer (parchment + PA/PE) system. The difference between those packaging variants is mainly based on different aerosol concentrations and interruption times between 'spraying' and 'packaging' steps.

Dipping the inside layer of PA/PE laminates in aloe gel immediately before cheese packaging may be another modification of the packaging method (variant IV). However, this variant may be risky because of possible discoloration of cheese surfaces and sizable costs.

A mathematical model may be used for the evaluation of packing methods and the usefulness of the new system. In particular, a stability evaluation model for such a packaging system would take the following form:

Table 1.7 Influence of storage time, bioaerosol addition and number of *Lactococcus* spp. bacteria on behaviour of staphylococci population in stored cheeses

Product type	Mathematical model form
Cheese stored without bioaerosol addition	$S = 377.034 - 248.149 \times t - 8.123 \times 10^{-5} \times L + 11.44 \times t^2 + 1.755 \times 10^{-5} \times t \times L - 1.059 \times 10^{-12} \times L^2$
Cheese stored with bioaerosol addition	$S_a = 1202.67 - 232.937 \times t - 5.122 \times 10^{-5} \times L_a + 11.263 \times t^2 + 1.223 \times 10^{-5} \times t \times L_a - 1.014 \times 10^{-12} \times L_a^2$

S_a *Staphylococcus aureus* in cheeses with addition of aloe, *S S. aureus in* cheese without aloe addition; L_a *Lactococcus* spp. in cheeses with addition of aloe, *L Lactococcus* spp. in cheeses without aloe addition, *t* storage duration

$$P = 0.5B(t) + 0.4A(t) + 0.1T(t) \tag{1.7}$$

where t corresponds to days, and:

- B is a factor describing barrier properties
- A concerns antibacterial properties
- T is a factor related to the influence of storage times
- a (coefficient of pertinence of barrier qualities) = 0.5
- b (coefficient of pertinence of antibacterial qualities) = 0.4
- c (coefficient of pertinence of storage time) = 0.1.

The presented equation is a simplified version of the model suggested by us for evaluation of packaging stability (Morawska et al. 2013). Another modification of packaging materials for lactic acid cheeses is a change in the material structure. The most perfect solution for this problem would be the possible incorporation of aloe into the PA/PE laminate. In other terms, powdered aloe gel should be applied on the PE matrix. The concentration of aloe powder should not exceed 5% due to its high water absorption (Variant V). Usage of low-density PE materials with thickness increased to 100 μm, modified by addition of dried aloe gel may be a good addition to the new PA/PE laminate.

Further research of qualities of such packaging will allow for the future elimination of economically adverse double-packaging systems, and usage of unsafe oxygen absorbents.

References

Fitzgerald M, Papkovsky DB, Smiddy M, Kerry JP, O'Sullivan CK, Buckley DJ, Guilbault GG (2001) Nondestructive monitoring of oxygen profile in packaged foods using phase-fluorimetric oxygen sensor. J Food Sci 66(1):105–110. doi:10.1111/j.1365-2621.2001. tb15590.x

Jasinska M, Harabin K, Dymytrów I (2014) Effect of packaging and season of milk production on selected quality characteristics of organic acid curd cheese during storage. Acta Sci Pol Technol Aliment 13(3):231–242. doi:10.17306/J.AFS.2014.3.1

Mazur J, Andrejko D, Masłowski A (2011) Wpływ zastosowanego podciśnienia w trakcie pakowania na podstawowe właściwości fizyczne serów twarogowych. Inżyn Roln 5(130): 179–184

Mazur J, Sobczak P, Panasiewicz M, Zawiślak K, Nieścioruk K, Wyrykowski G, Żak W (2012) Wpływ ciśnienia pakowania twarogów kwasowych na wybrane parametry produktu. Inżyn Roln 30(138):139–146

Morawska M (2012) Wpływ komponentów twarogów na właściwości laminatów poliamidowo/ polietylenowych. Dissertation, Gdynia Maritime University

Morawska M, Steinka I, Blokus-Roszkowska A (2013) Modelowanie matematyczne w ocenie jakości materiałów opakowaniowych. Zeszyty Naukowe Akademii Morskiej w Gdyni 80:5–12

Panfil-Kuncewicz H, Staniewski B, Szpendowski J, Nowak H (2006) Application of active packaging to improve the shelf live of fresh white cheeses. Pol J Food Nutr Sci 15(16): 165–168. doi:10.15193/zntj/2014/93/190-203

Parisi S (2016) The world of foods and beverages today. Online video course, Learning.ly/The Economist Group. Available http://learning.ly/

Steinka I (2003) Wpływ interakcji opakowanie produkt na jakość mikrobiologiczna hermetycznie pakowanych serów twarogowych. Wydawnictwo Akademii Morskiej, Gdynia, pp 1–111

Steinka I (2005a) Evaluation of changes in properties of PA/PE film used for vacuum-packed lactic acid cheese. Iran Polym J 14(1):5–13

Steinka I (2005b) Wpływ jakości mikrobiologicznej twarogów na fazę wodną w warunkach pakowania hermetycznego. Roczniki PZH 56(3):275–281

Steinka I (2005c) Wyznaczanie jakości higienicznej kwasowych serów twarogowych wytwarzanych w różnych warunkach. Bromatologia i Chemia Toksykologiczna Suppl, pp 373–376

Steinka I (2006) Możliwość zastosowania programu komputerowego do prognozowania wpływu czynników mikrobiologicznych na właściwości opakowań. Ochrona przed korozją 9s/A:93–96

Steinka I (2007a) Wpływ czynników mikrobiologicznych na masę laminatów stosowanych do pakowania twarogów. Roczniki PZH 58(2):459–469

Steinka I (2007b) Sposób antygronkowcowej suplementacji skrzepu twarogowego. Biuletyn Urzędu Patentowego Rzeczpospolita Polska, Patent number 197169

Steinka I (2007c) The influence of interactions occurring between micro-organisms on predicting the safety of lactic acid cheese. In: PletneyVN (ed) Focus on food engineering research and developments. Nova Science Publishers, Inc., New York, pp 165–237

Steinka I (2008a) Lactic acid cheese safety. Nova Science Publishers, Inc., New York

Steinka I (2008b) Modelowanie przeżywalności psychrotrofów w twarogach pakowanych hermetycznie. Bromatologia i Chemia Toksykologiczna 41(3):57–64

Steinka I (2008c) Biofilm formed on the surface of PA/PE laminates packaging. Proceedings of the biofilms III: 3rd international conference, Munich, 6–8 Oct 2008, p 86

Steinka I (2009a) Assessment of interaction occurring between micro-flora and packaging applied for food. In: Bellinghouse VC (ed) Food Processing: methods, techniques and trends. Nova Science Publishers Inc, pp 463–491

Steinka I (2009b) The influence of the interactions between lactic acid cheese micro-organisms on predict the properties of packaging. Joint Proc Wydawnictwo Akademii Morskiej Gdynia-Hochschule Bremenhaven 22:12–18

Steinka I (2011) Zmiany jakości serów twarogowych dostępnych w sieciach handlowych w okresie ostatnich kilkunastu lat. Przegląd Mleczarski 11:14–18

Steinka I (2012) Prognozowanie mikrobiologiczne jako narzędzie oceny jakości kwasowych serów niedojrzewających. Przegląd Mleczarski 4:10–14

Steinka I, Blokus-Roszkowska A (2009) Application of tertiary mathematical models for evaluating the presence staphylococcal enterotoxin in lactic acid cheese. In: Martorell S, GuedesSoares C, Barnett J (eds) Safety, reliability and risk analysis: theory, methods and applications, vol 1. Taylor & Francis Group, London, pp 2269–2273

Steinka I, Kukulowicz A (2003) Próba optymalizacji jakości twarogów za pomocą aerozolu aloesowego. Bromatologia i Chemia Toksykologiczna 4:341–346

Steinka I, Kukulowicz A (2004a) Assessement of adherence degree of the *Lactococcus sp.* to the surface of PA/PE laminates. Joint Proc Wydawnictwo Akademii Morskiej, Gdynia-Hochschule Bremenhaven 17:44–50

Steinka I, Kukulowicz A (2004b) Adhesion of the *Lactococcus sp.* to surface of traditional and biodegradable packaging. Pol J Nat Sci 2:151–165

Steinka I, Morawska M (2006) Zmiany właściwości fizycznych laminatów poliamidowo/polietylenowego pod wpływem działania serwatki i skrzepu bakterii fermentacji mlekowej. Ochrona przed korozją 9s/A:89–91

Steinka I, Morawska M (2007) Lactic acid solution influence on polyamide/polyethylene laminates properties. Ann Pol Chem Soc 549–552

Steinka I, Parisi S (2006) The influence of cottage cheese manufacturing technology and packing metod on the behaviour of micro-flora. Joint Procs Wydawnictwo Akademii Morskiej Gdynia-Hochschule Bremenhaven 19:30–37

Steinka I, Stankiewicz J (2002) Wpływ czynników związanych z opakowaniami na wybór twarogów przez konsumentów. Materiały Naukowe XXIII Sesji Naukowej KTiCHŻ PAN Lublin, p 355

Steinka I, Morawska M, Kukulowicz A (2011) The influence of lactic acid cheese components on the selected properties of polyamide/polyethylene laminates. Curr Trends Commod Sci, Wydawnictwo Uniwersytetu Ekonomicznego Poznań 186:56–66

Chapter 2
Evolutive Profiles of Caseins and Degraded Proteins in Industrial Cow's Milk Curds

Caterina Barone, Marcella Barbera, Michele Barone, Salvatore Parisi and Izabela Steinka

Abstract The importance of prepackaged curds in the current cheese market is increased in the last years because of the persistence of cyclic periods with remarkable diminution of stored raw materials. Consequently, the cyclic deficiency of cow's milk may determine the subsequent lack of correlated derivatives and force manufacturers to use prepackaged curds. Because of the critical importance of the chemical and microbiological 'quality' of these curds, the study of evolutive profiles of casein contents in selected industrial curds should be recommended. The aim of this chapter has been to show the analytical results of an industrial study carried out on seven different cow's milk curds during storage. Obtained data and calculated results seem to suggest that curd samples under refrigerated conditions can show increased proteolysis. In addition, moisture and pH values may show notable augments during refrigerated storage. On the other side, deep-frozen storage is recommended when speaking of curd use after extended storage times.

Keywords Casein · Cow's milk curd · CYPEP:2006 · Frozen storage · Lipids · Moisture · Refrigerated storage

C. Barone (✉)
Associazione "Componiamo il Futuro" (COIF) Palermo, Palermo, Italy
e-mail: fct1970@libero.it

M. Barbera
DEMETRA Department, University of Palermo, Palermo, Italy
e-mail: marcellapu@hotmail.it

M. Barone
Associazione "Componiamo il Futuro" (COIF) Palermo, Palermo, Italy
e-mail: fct1970@libero.it

S. Parisi
Industrial Consultant, FSPCA PCQI, Palermo, Italy
e-mail: drparisi@inwind.it

I. Steinka
Gdynia Maritime University, Gdynia, Poland
e-mail: ize13ste@wp.pl

© The Author(s) 2017
C. Barone et al., *Chemical Profiles of Industrial Cow's Milk Curds*,
Chemistry of Foods, DOI 10.1007/978-3-319-50942-6_2

Abbreviations

CYPEP:2006   Cheesemaking Yield and Proteins Estimation according to
              Parisi:2006
CUR   Curd sample
FC   Fat matter
MC   Moisture
PC   Proteins

2.1 Introduction

The importance of prepackaged cow's milk curds in the current market of cheeses is
increased in the last years because of the persistence of cyclic periods with
remarkable diminution of stored raw materials for cheesemaking purposes.
Actually, these 'pauses' have determined the overproduction of cheeses and other
dairy foods in certain periods of the year (Barbieri et al. 2014a). The current
economic crisis on a global scale is also important and should be taken into account
(Parisi 2016): as a result, the cyclic deficiency of cow's milk may determine the
subsequent lack of correlated derivatives (butters, caseins, yoghurts, etc.).

With exclusive reference to cow's milk-made cheeses—other milks have certainly their own importance (Alais 1984), one of the most critical factors in the
production of these foods is the good condition (also named 'quality' by cheesemakers) of the initial milk, in terms of microbiological counts and profiles on the
one hand, and chemical–physical features on the other side (Pacheco and Galindo
2010; Zagare et al. 2014).

As a consequence, 'good' milk can be used to produce good cheeses, on condition that initial conditions are acceptable or excellent. However, the production of
normal cheeses cannot be performed without a necessary (and crucial) step: the
production of the intermediate 'curd'. This step is not described in detail when
speaking of cheeses and cheese-like products obtained with complex formulations
such as 'low moisture' mozzarella cheeses (Banks 2007; Barbieri et al. 2014a). By
the chemical viewpoint, curd is the coagulated and precipitated heterogeneous
matter from the original milk (Alais 1984).

In general, the composition of similar intermediates concerns five main components (Barbieri et al. 2014a; McSweeney 2007a):

- Water (curds are substantially solid solutions)
- Lipids, chemically defined as esters, derived from glycerol and three fatty acids.
  These large molecules are trapped into the curd agglomeration by means of a
  protein matrix (Parisi and Caruso 2013)
- Proteins, casein molecules above all. These coagulated molecules can trap fats
  and other compounds into the precipitated curd

- Mineral salts, generally calcium salts. Added sodium chloride may be important depending on the peculiar cheese (McSweeney 2007b)
- Carbohydrates in small quantities.

Basically, each parameter has an important role when speaking of initial milk and the correlated cow's milk curd, with the exclusion of carbohydrates and mineral salts, although calcium and phosphate compounds are important when speaking of caseins. In general, it can be affirmed that:

(a) Water is the liquid medium for curds and original milks. With relation to cheese production, the higher the amount of water in curds, the lower the incorporation of additional water molecules during the next steps. However, because of the remarkable variety and differentiation of cheese types, the increase of product weights may be not important. For example, semi-hard and hard cheeses have to naturally contain small aqueous amounts

(b) The amount of lipids is important when speaking of cheese palatability. In general, the more abundant the fraction of fat matter on the dry content, the higher the perceived sweet taste in soft and unripened cheeses. On the other hand, certain products with prevailing microbial fermentation can exhibit peculiar flavours and tastes depending on the action of selected microorganisms on lipids. In fact, lipolytic life forms such as yeasts and lactobacilli can produce volatile substances and other degradation products with interesting features by the consumers' viewpoint (Barbieri et al. 1994; Centeno et al. 1996; Delgado et al. 2016; Fox and Wallace 1997; Medina et al. 1995; Reiter et al. 1969)

(c) Proteins are the real key factor in milk coagulation. Actually, the most part of these organic chains constituted of different caseins (Kelly 2007; Parisi 2006; Parisi et al. 2006a, 2009). After the formation of a protein agglomeration around a hypothetical centre (Barbieri et al. 2014a; Fox and McSweeney 1998; Guinee 2007) and the incorporation of lipids, carbohydrates, mineral salts and water, caseins may also be linked to additional water molecules by means of hydrogen bonds. On the other side, this absorption can be limited depending on the condition of lipids, the composition of partially demolished proteins, and the availability of binding calcium ions (these particles are present in the original protein as phosphate salts).

On these bases, it may be also remembered that chemical–physical parameters for the intermediate curd and the final product depend on the chemical characterisation of lipids, water and proteins. As a result, analytical determinations such as pH and acidity are notably influenced if lipids are partially decomposed, water is excessively incorporated, and/or caseins are fragmented, with or without calcium ions. It should be mentioned also that additional water amounts are continually generated by hydrolysis because of microbial activity and other chemical reactions. Consequently, several unripened and/or packaged soft cheeses products appear to increase aqueous amounts, pH and redox potential values after some storage months, depending also on storage conditions (Parisi 2002, 2003).

As a result, the composition of caseins has a remarkable and critical influence on properties of final cheeses (Barbieri et al. 2014a), including good textural properties. In general, high yield values and good texture in cheese production are linked with the amount and the average molecular weight of caseins above all. Consequently, an interesting field of research is the study of evolutive profiles of selected molecules and protein aggregations in cheeses. When speaking of nitrogen-based molecules, the analytical research is normally carried out by means of direct experimental protocols such as Kjeldahl and Dumas methods (Alais 1984). However, the study of proteolysis in certain cheeses has been already tried by means of an indirect method, the 'Cheesemaking Yield and Proteins Estimation according to Parisi: 2006' (CYPEP:2006) approach (Parisi et al. 2006b; 2016a,b). This mathematical and simulative procedure has been used with the aim of evaluating potential differences between industrial curds.

The aim of this Chapter is the possibility of tracking evolutive profiles of casein contents in selected industrial curds for subsequent cheese production. An industrial study has been carried out on seven different cow's milk curds near a cheesemaking company (none of examined curds has been produced by this industry). General differences between these curds (prepackaged products) concern the chemical composition, pH values, and the possibility of prolonged frozen storage instead of refrigerated conditions.

2.2 Evolutive Profiles of Selected Cow's Milk Curds During Storage

2.2.1 Materials

Seven different productions of cow's milk curds have been considered for this study. In detail, five lots by different curd Producers have been sampled at the arrival near a cheesemaking industry under refrigerated storage (delivery temperature: 2 ± 2 °C; two 200.0 g—samples per lot) and subsequently re-sampled after eight and 15 days of storage (temperature: ≤ 10 °C; two 200.0 g—samples per lot). As a result, six samples have been obtained for each curd production. The remaining curds have been received and sampled under refrigerated storage (delivery temperature: 2 ± 2 °C; two 200.0 g—samples per lot); however, subsequent samples have been obtained after eight and 15 days of deep-frozen storage (temperature: ≤ -18 °C; two 200.0 g—samples per lot).

All received curds have been found to be vacuum-packaged as 25–50 kg—blocks with thermosealable films, generally polypropylene or polyamide/polyethylene plastic matters.

As a result, 42 total samples have been considered and stored at $2° \pm 2$ °C before analyses. With relation to this study, one single curd sample (CUR) has been named 'CUR xxx' ('xxx' is an acronym used for the representation of the external lot and producer) and immediately analysed after 24 h (two samples per lot, two analyses). The remaining two sample groups (four samples: two sampled after eight days; two sampled after 15 days) have been named 'CUR xxx-a' ('a' means the number of days after the arrival) and analysed after eight and 15 days. For instance, the second sample group for curd CUR 001 has been named 'CUR 001-8' and analysed after 8 days.

All sampled cheeses have been analysed according to the above mentioned schedule. Moisture, fat Matter, pH, and proteins have been evaluated for all sampled products.

2.2.2   Analytical Methods

Moisture (MC), fat matter (FC), and pH have been obtained for all sampled products with the following methods respectively: IR thermogravimetric method; AFNOR NF V04-287 (Barbieri et al. 2014b); pH Meter method. The most reliable amount of proteins (PC) has been calculated by means of the 'Compact Cheese Spreadsheets' software, version 1.1, CYPEP Lite.bas, and CheeSim$.bas (Parisi et al. 2016a, b). These software are all based on CYPEP:2006 indirect method. All results have been obtained as the average of two data per sample.

2.2.3   Results and Discussion

Average data for sampled cheeses (chemical results) are displayed in Table 2.1 in function of days after the arrival and analysis (one, eight, and 15 days). Displayed data correspond to the average value of the whole group of samples for MC, FC, pH and PC respectively. In addition, Table 2.2 shows pH and PC average data for two different subgroups:

Table 2.1 Chemical data for stored curds, average data

Storage days	Stored curds, refrigerated and deep-frozen conditions (average data)			
	MC, %	FC, %	pH	PC, %
1	43.8	27.1	5.27	24.5
8	44.9	26.3	5.33	24.2
15	45.8	25.9	5.37	23.8

MC is for: moisture, FC is for: fat matter, PC represents proteins (the most reliable amount for proteins, according to CYPEP:2006). The most reliable amount of proteins (PC) has been calculated by means of the 'Compact Cheese Spreadsheets' software, version 1.1, CYPEP Lite. bas, and CheeSim$.bas. These software are all based on CYPEP:2006 indirect method

Table 2.2 pH and protein values for stored curds under refrigerated conditions, average data

	Stored curds, refrigerated conditions (average data)								
	CUR 001		CUR 002		CUR 003		CUR 004		CUR 005	
Storage days	pH	PC, %	pH	PC, %	pH	PC, %	pH	PC, %	pH	PC; %
1	5.22	25.3	5.38	23.6	5.20	25.2	5.29	23.9	5.19	25.2
8	5.28	25.2	5.42	23.0	5.30	24.8	5.32	23.0	5.30	24.2
15	5.30	24.2	5.50	22.7	5.34	24.5	5.44	22.7	5.37	24.0

PC represents proteins (the most reliable amount for proteins, according to CYPEP:2006. PC has been calculated by means of the 'Compact Cheese Spreadsheets' software, version 1.1, CYPEP Lite.bas, and CheeSim$.bas. These software are all based on CYPEP:2006 indirect method

- CUR 001-005. This group concerns five curds stored under refrigerated conditions until 15 days
- CUR 005-006. This group concerns two curds stored under deep-frozen conditions until 15 days.

In general, initial sampled curds show a variegated composition when speaking of moisture, fat matter, pH, and proteins. MC is between 42.0 and 45.9% (average value: 43.8%, Table 2.1); FC ranges from 26.0 to 28.0% (average value: 27.1%); pH values are between 5.19 and 5.38 (medium calculated result: 5.27). As a consequence, calculated amounts for PC give a general average value of 24.5% (minimum value: 23.9%; maximum value: 25.3%, Tables 2.2 and 2.3).

After refrigerated and deep-frozen storage, results show a notable increase of MC and pH values, while FC and PC amounts tend to lower values (Table 2.1). In detail, MC increases from 43.8 to 44.9% after eight days (augment: +1.1%) and 45.8% after 15 days (augment: +2.0%). The same trend is apparently shown by pH values (+0.10 after 15 days), while FC decreases (difference: −1.2% after 15 days). Consequently, PC profiles seem to show a clear diminution: −0.3 and −0.7% after eight and 15 days respectively (Table 2.1).

Table 2.3 pH and protein values for stored curds under deep-frozen conditions, average data

	Stored curds, deep-frozen conditions (average data)			
	CUR 006		CUR 007	
Storage days	pH	PC, %	pH	PC, %
1	5.22	25.3	5.38	23.6
8	5.27	25.1	5.42	23.6
15	5.25	25.2	5.40	23.4

PC represents proteins (the most reliable amount for proteins, according to CYPEP:2006. PC has been calculated by means of the 'Compact Cheese Spreadsheets' software, version 1.1, CYPEP Lite.bas, and CheeSim$.bas. These software are all based on CYPEP:2006 indirect method

However, two sample curds have been stored under deep-frozen conditions (Table 2.3) differently from the first five samples (Table 2.2). Obtained and calculated results show that:

(a) pH increase and PC diminution are strictly correlated during the 15 days storage in both situations, when speaking of average data
(b) Deep-frozen storage (Table 2.3) seems to slow down the degradation of nitrogen-based molecules (proteins), because the observed diminution is only −0.2%. On the other side, pH values (average data) appear to be increased after 15 days. Probably, experimental analytical bias may have some influence
(c) Refrigerated curds (Table 2.2) show a very notable increase of pH values: +0.13 after 15 days. On the other side, PC values decrease from 24.6 (average data on five samples) to 24.0 and 23.6% after eight and 15 days respectively. This trend confirms the general situation of Table 2.1.

2.3 Conclusions

Obtained data and calculated results seem to suggest that curd samples under refrigerated conditions can show increased proteolysis (Parisi et al. 2004), while lipolysis may be observed with minor importance. In addition, moisture increases notably (as the result of enhanced proteolysis) and pH values grow up during storage. On the other side, deep-frozen storage is recommended when speaking of curd use after extended storage times. Actually, protein profiles seem to decrease slightly even in these conditions, but observed differences may be also ascribed to experimental and analytical bias.

References

Alais C (1984) Science du lait. Principles des techniques laitières, 4th edn. S.E.P.A.I.C., Paris
Banks JM (2007) How is cheese yield defined? In: McSweeney PLH (ed) Cheese problems solved. Woodhead Publishing Limited, Cambridge, and CRC Press LLC, Boca Raton, pp 102–104
Barbieri G, Bolzoni L, Careri M, Mangia A, Parolari G, Spagnoli S, Virgili R (1994) Study of the volatile fraction of Parmesan cheese. J Agric Food Chem 42(5):1170–1176. doi:10.1021/jf00041a023
Barbieri G, Barone C, Bhagat A, Caruso G, Conley ZR, Parisi S (2014a) The problem of aqueous absorption in processed cheeses: a simulated approach. In: Barbieri G, Barone C, Bhagat A, Caruso G, Conley ZR, Parisi S (eds) The influence of chemistry on new foods and traditional products. Springer International Publishing, pp 1–17
Barbieri G, Barone C, Bhagat A, Caruso G, Conley ZR, Parisi S (2014b) The prediction of shelf life values in function of the chemical composition in soft cheeses. In: Barbieri G, Barone C, Bhagat A, Caruso G, Conley ZR, Parisi S (eds) The influence of chemistry on new foods and traditional products. Springer International Publishing, Heidelberg

Centeno JA, Menéndez S, Rodriguez-Otero JL (1996) Main microbial flora present as natural starters in Cebreiro raw cow's-milk cheese (Northwest Spain). Int J Food Microbiol 33(2):307–313. doi:10.1016/0168-1605(96)01165-8

Delgado AM, Parisi S, Almeida MDV (2016) Milk and dairy products. In: Delgado AM, Almeida MDV, Parisi S (eds) Chemistry of the Mediterranean diet. Springer International Publishing, Cham, pp 139–176

Fox PF, McSweeney PLH (1998) Dairy chemistry and biochemistry. Blackie Academic and Professional, London

Fox PF, Wallace JM (1997) Formation of flavor compounds in cheese. In: Neidleman SL, Laskin AI (eds) Advances in applied microbiology, vol 45. Academic Press, New York and London, pp 17–86

Guinee TP (2007) What effects does cold storage have on the properties of milk? In: McSweeney PLH (ed) Cheese problems solved. Woodhead Publishing Limited, Cambridge, and CRC Press LLC, Boca Raton, pp 28–29

Kelly AL (2007) Milk. What is the typical composition of cow's milk and what milk constituents favour cheesemaking? In: McSweeney PLH (ed) Cheese problems solved. Woodhead Publishing Limited, Cambridge, and CRC Press LLC, Boca Raton, pp 3–4

McSweeney PLH (2007a) Introduction: how does rennet coagulate milk? In: McSweeney PLH (ed) Cheese problems solved. Woodhead Publishing Limited, Cambridge, and CRC Press LLC, Boca Raton, pp 50–51

McSweeney PLH (2007b) Milk. Introduction. In: McSweeney PLH (ed) Cheese problems solved. Woodhead Publishing Limited, Cambridge, and CRC Press LLC, Boca Raton, pp 1–2

Medina MRL, Tornadijo ME, Carballo J, Sarmiento RM (1995) Mi-crobiological study of León raw cow-milk cheese, a Spanish craft variety. J Food Prot 58(9):998–1006

Pacheco FP, Galindo AB (2010) Microbial safety of raw milk cheeses traditionally made at a pH below 4.7 and with other hurdles limiting pathogens growth. In: Méndez-Vilas A (ed) Current research technology education topics applied microbiology microbial biotechnology, FORMATEX Microbiology Series No. 2, pp 1205–1216. Available http://www.formatex.info/microbiology2/1205-1216.pdf. Accessed 04 Oct 2016

Parisi S (2002) Profili evolutivi dei contenuti batterici e chimico-fisici in prodotti lattiero-caseari. Ind Aliment 41(412):295–306

Parisi S (2003) Evoluzione chimico-fisica e microbiologica nella conserva-zione di prodotti lattiero - caseari. Ind Aliment 42(423):249–259

Parisi S (2006) Profili chimici delle caseine presamiche alimentari. Ind Aliment 45(457):377–383

Parisi S (2016) The world of foods and beverages today. Online video course, Learning.ly/The Economist Group, 750 Third Ave NY, NY 10017. Available http://learning.ly/products/the-world-of-foods-and-beverages-today-globalization-crisis-management-and-future-perspectives

Parisi S, Caruso G (2013) Il regolamento CE 2073/2005 e successivi aggior-namenti. Studio di piani di campionamento ridotti per formaggi molli altamente deperibili. Ind Aliment 52 (541):8–17

Parisi S, Delia S, Laganà P (2004) Il calcolo della data di scadenza degli alimenti: la funzione Shelf Life e la propagazione degli errori sperimentali. Ind Aliment 43(438):735–749

Parisi S, Laganà P, Delia S (2006a) Curve di crescita dei miceti in diversi formaggi in atipiche condizioni di conservazione. Ind Aliment 45(458):532–538

Parisi S, Laganà P, Delia AS (2006b) Il calcolo indiretto del tenore proteico nei formaggi: il metodo CYPEP. Ind Aliment 45(462):997–1010

Parisi S, Laganà P, Stilo A, Micali M, Piccione D, Delia S (2009) Il mas-simo assorbimento idrico nei formaggi. Tripartizione del contenuto acquoso per mole d'azoto. Ind Aliment 48, 491:31–41

Parisi S, Barone C, Caruso G (2016a) La standardizzazione dei parametri chimici iniziali nei formaggi semiduri tramite l'uso di un programma in Basic: il CYPEP Lite, versione 1.00.1. Ind Alimen 55(568):3–6

Parisi S, Barone C, Caruso G (2016b) Packaging failures in frozen curds. The use of a BASIC software for mobile devices. Food Packag Bull 25(1 & 2):16–19

Reiter B, Sorokin Y, Pickering A, Hall AJ (1969) Hydrolysis of fat and protein in small cheeses made under aseptic conditions. J Dairy Res 36(1):65–76. doi:10.1017/S0022029900012541

Zagare MS, Ghorade IB, Patil SS (2014) Effect of pH on microbial quality of milk. Indian J Appl Res 4(5):272–274

Chapter 3
The Production of Industrial Cow's Milk Curds

Caterina Barone, Marcella Barbera, Michele Barone, Salvatore Parisi
and Izabela Steinka

Abstract By the historical viewpoint, cheeses are the final product of a complex work starting from milk. This liquid substance has to be coagulated: as a result, a heterogeneous matter—curd—is precipitated from the original milk. The rough composition of this intermediate influences the chemical composition of the final cheese. For these reasons, the production of industrial curds should be studied with attention. Different systems are available at present, but the mail process concerns always the coagulation of the original milk and the production of readily available curds for immediate or subsequent use (near the same cheesemaker or in different locations).This chapter discusses the normal production of prepackaged and ready-to-use cow's milk curds with several technological modifications for industrial purposes.

Keywords Acidification · Casein · Coagulation · Cooling · Cow's milk curd · Cutting · Draining · Rennet · Salt · Starter culture

C. Barone (✉)
Associazione "Componiamo il Futuro" (COIF) Palermo, Palermo, Italy
e-mail: fct1970@libero.it

M. Barbera
DEMETRA Department, University of Palermo, Palermo, Italy
e-mail: marcellapu@hotmail.it

M. Barone
Associazione "Componiamo il Futuro" (COIF) Palermo, Palermo, Italy
e-mail: fct1970@libero.it

S. Parisi
Industrial Consultant, FSPCA PCQI, Palermo, Italy
e-mail: drparisi@inwind.it

I. Steinka
Gdynia Maritime University, Gdynia, Poland
e-mail: ize13ste@wp.pl

© The Author(s) 2017
C. Barone et al., *Chemical Profiles of Industrial Cow's Milk Curds*,
Chemistry of Foods, DOI 10.1007/978-3-319-50942-6_3

3.1 Introduction

At present, the market of industrial cheeses seems to be notably influenced on a large scale by the 'global' availability of milk and derived products. Actually, the above-mentioned situation has been always observed on a small or medium scale when speaking of restricted economic areas (Parisi 2016), but the increasing dimension of global food markets has certainly obliged cheese manufacturers in several 'unlucky' countries to purchase remarkable amounts of milks for their productions (Alais 1984; Parisi et al. 2016).

In general, the ready availability of certain raw materials may force producers, packagers, and retailers to 'impose' new or ameliorated versions of existing and historical traditions to consumers. For this basic reason, it may be inferred that the influence of globalisation is extremely important at present, and the future could demonstrate the correctness of this affirmation. However, food producers have to comply with different implicit and explicit requests by several 'stakeholders', including regulatory agencies, mass retailers, and 'the final consumer'. Many aspects should be taken into account when speaking of the 'design' of new or current foods in the modern world (Parisi 2016).

In addition, readily available raw materials do not automatically mean 'excellent' or 'acceptable' quality with relation to purchased commodities. In other words, food raw materials are subjected to degradation, microbial spreading, packaging damages, and other transformations of the original product before of the final use by food producers (Parisi et al. 2016). As a clear consequence, the simple delivery of edible raw materials concerns several management approaches, and this aspect has to be carefully considered.

With relation to cheeses, different milk types can be used. In general, cow's milk seems to be the most important ingredient for the production of many cheeses worldwide (Banks 2007a), although buffalo's or goat's milk are extremely appreciated (Barbieri et al. 2014).

With relation to cow's milk cheeses, different classifications have been proposed so far. In general, it may be assumed that all systems appear to be based mainly on the technology of production, a few chemical–physical features, and the microbial population of the final cheese (Barbieri et al. 2014; Burkhalter 1981; Codex Alimentarius Commission 1978, 1999; McSweeney 2007; McSweeney et al. 2004; Ottogalli 2001). Anyway, it has to be clarified that similar classifications include also processed (also named melted or cheese-like or analogous) cheeses. The last heterogeneous group of foods concerns all productions with more than one basic ingredient: the original milk (Barbieri et al. 2014).

By the historical viewpoint, cheeses are the final product of a complex work starting from milk. This liquid substance has to be coagulated: as a result, a heterogeneous matter is precipitated from the original milk. The rough composition of this intermediate influences the chemical composition of the final cheese (Kelly 2007a). Five main components—water, carbohydrates, proteins, lipids, and mineral salts—are the basic constituents of the obtained curd (Sect. 2.1).

Because of the possible scarcity of readily available milk worldwide, the market of milk and cheeses has progressively included intermediates for different productions: animal butters, casein products, lactose, etc. This category includes also 'industrial curds': a well-defined category of curds obtained by milk coagulation and immediately packaged with the aim of supplying cheesemaking industries with limited use of transport trucks. In fact, the delivery of a 25,000 kg—amount of prepackaged curds is economically more convenient than the delivery of an equivalent amount of the original milk. As an example, 25,000 kg of prepackaged curd correspond to 250,000 l of initial milk at least; consequently, the volumetric capacity of delivery trucks is notably reduced.

In addition, untreated, pasteurised, or thermally treated milks can always suffer severe alterations from the chemical and microbiological viewpoints before use near the final producer. As a result, the utilisation of preproduced curds makes easier the subsequent production of safe cheeses.

For all these reasons, the production of industrial curds should be studied with attention. Different systems are available at present, and the use of a peculiar coagulation method is generally correlated with several of chemical and physical features of these food products. Actually, industrial curds have to be considered as raw materials for the production of different cheeses; consequently, the higher the range of available curds (type of milk, pH values, microbial fermentation, colour, etc.), the higher the number of hypothetically available cheeses.

3.2 The Production of Industrial Curds. A Practical Description

The ingredient list of industrial (ready-to-use) curds generally shows the following ingredients, although several exceptions may occur:

(1) Fresh, pasteurised, or thermally treated cow's milk
(2) Salt (sodium chloride)
(3) Rennet (animal, vegetable, or microbial origin)
(4) Starter culture.

The order of these ingredients may also be modified because of the position of 'starter culture' (lactobacilli, etc.).

Similarly to normal cheeses, curds can also be discriminated on the basis of the production process (example: use of microbial, vegetable, or animal rennet) or chemical–physical features. The most important of these data appears to be the 'fat matter on dry content' amount because higher numbers mean the possibility of final cheeses with strong flavours, sweet or 'fatty' tastes (depending on the microbial ecology of the product). In addition, different curds mean always different 'expiration dates' for food manufacturers. For this reason, several types are sold, delivered and stored under deep-frozen conditions instead of refrigerated storage with the aim of extending these expiration dates.

The process of production for these ready-to-use curds has to be considered as the first step of the entire production chain for the final product. Naturally, cheddar or mozzarella cheeses require different curds with concern to moisture, fat matter on dry content, microbial counts, and so on. The aim of this chapter is to give a simplified description of normal production processes for industrial curds.

3.2.1   Industrial Curds. Production Steps

In general, the following steps should be considered when speaking of curd production, although the exact sequence and several conditions—examples: times, temperatures, etc.—may vary (Anonymous 2016):

(a) Milk standardisation
(b) Milk treatments (pasteurisation or different heat treatments)
(c) Milk cooling
(d) Addition of starter culture
(e) Optional step: addition of nonstarter microorganisms
(f) Curd ripening
(g) Rennet coagulation
(h) Curd cutting and heat treatment
(i) Curd draining.

Once more, it has to be clarified that this process is only one of the many possibilities into cheese/curd production industries. In addition, the curd production is only a series of steps into the more general and complete flowchart of the final product (cheese).

The following Sections give a simplified description of each step.

3.2.1.1   Milk Standardisation

Different milks are usually mixed with the aim of obtaining constant chemical–physical features in the final curd. For this reason, initial analyses aim to determine protein and fat amounts in the original milks.

3.2.1.2   Milk Treatments

Milks used for curd (and cheese) productions may be used as 'fresh' or pasteurised liquids, although other different thermal treatments can be used. It should be considered that:

(a) Thermal treatments such as pasteurisation at 73 °C for 15 s at least are requested and carried out for safety purposes
(b) The pasteurisation and other heat treatments may be very helpful if the initial microflora has to be destroyed with the aim of promoting the positive development of 'starter culture' (Sect. 3.2.1.4).

3.2.1.3 Milk Cooling

Should initial milks be heat treated, the development of starter culture would be very difficult at the end of heating processes. For this reason, a thermal value of 32 °C should be assured by means of a specific cooling step.

3.2.1.4 Addition of Starter Culture

Peculiar starter bacteria—lactobacilli, streptococci—are needed because of two factors:

(1) pH of the liquid medium (milk) has to be lowered before rennet coagulation. Because of the attitude of starter culture to produce organic acids by fermentation, the objective is reached in a 'natural' and 'traditional' way. Acid pH values are required in the coagulation step (enzymes work well in acid conditions)
(2) Some peculiar lactobacilli may be used with the aim of obtaining peculiar texture, flavours and/or tastes in final cheeses. Consequently, these microorganisms cannot be defined 'starter culture' if the declared aim of starters is only linked to pH lowering. However, these bacteria are also lactobacilli. More precisely, these 'enhancer agents' have different duties, by the technological viewpoint (Sect. 3.2.1.5).

Times are extremely variable: generally, this step should be carried out at 32 °C and a total time of 150 min might be needed when speaking of milk ferment addition (Ferrari et al. 2003).

3.2.1.5 Optional Step: Addition of 'Enhancer' Microorganisms

Certain life forms—some yeast, propionic bacteria, and peculiar lactobacilli (Sect. 3.2.1.4) may be added at this step if needed, depending on the final destination of curds.

3.2.1.6 Curd Ripening

Added starter culture can grow up their number during this step. pH values decrease and microbial fermentation processes are also responsible for the aroma of final products (in certain situations, sensorial features depend on the peculiar starter and/or 'enhancer' microorganism).

3.2.1.7   Rennet Coagulation

Curd is the product of an enzymatic coagulation of caseous matters into the original milk. Rennet enzymes—calf rennet, chymosin (extracted by calf rennet), vegetable or microbial proteinases, etc.—can act for a specified time: about 30 min, although times strongly depend on the original state of milks (Anonymous 2016); as a result, a casein-made matrix is able to precipitate with incorporated fat matters, some residual carbohydrate, mineral salts, and water. One of casein molecules, κ-casein, is attached by coagulating enzymes such as chymosin, and a glycomacropeptide fraction is removed with the consequent precipitation of κ-casein (Banks 2007a; Barbieri et al. 2014; O'Mahony et al. 2005) with the important role of calcium ions as chelating agent between different protein chains.

3.2.1.8   Curd Cutting and Heat Treatment

The coagulation process must continue until a desired pH value is reached. Ideally, pH should be 6.4 (range: 6.3–6.6) when speaking of rennet coagulation only (Belitz et al. 2009), while lactic acidification processes from sour milk should arrive to the isoelectric point of caseins, at pH 4.6 (range: 4.6–4.9). In particular, the action of coagulating enzymes such as chymosin is favoured at low pH values. In addition, the intermediate mass has to be cut with the aim of increasing the superficial area and allowing the expulsion of a significant amount of absorbed water in excess, including serum proteins also. A heat processing (temperature about 38 °C) can be performed with the same objective (Anonymous 2016), although this strategy is also helpful when speaking of control of undesired psychotropic life forms (Beresford 2007). The total duration of the entire step may vary depending on situations.

3.2.1.9   Curd Draining

The so-called 'whey' is removed as explained in Sect. 3.2.1.8. The final matter is a heterogeneous casein-made matrix with incorporated fat matter, water, and mineral salts. This is the 'final' step of curd procedures.

However, the standardisation of curd productions and the increasing automation into cheese companies has modified slightly the historical traditions. The following Section discusses some of the most important modifications to the general process.

3.2.1.10 Curd Production. Industrial Modifications

The duration of milk acidification may be notably reduced by means of the addition of different acids: hydrochloridric acid (low temperature), citric, or lactic acid, etc. (Alais 1984). This chemical approach allows to reach the desired pH value with reduced times if compared to 90–150 min with addition of milk ferment (Anonymous 2016). Anyway, pH values have to be fixed between 4.5 and 4.7, and the consequent flocculation (this process is not a real coagulation) is not obtained immediately (low temperature: about 5 °C), but in a subsequent step, at about 37 °C (Alais 1984).

The addition of calcium chloride in the original milk can be important enough because of ameliorated coagulation speed (reduced times) and higher firmness. This salt is normally added with coagulant enzymes (Barbieri et al. 2014; Parisi 2006).

Heat treatments may be performed during the production with the aim of recovering a certain quantity of denatured whey proteins into the curd.

The ultrafiltration process can be also used with a similar objective: the forced incorporation of native whey proteins (Kelly 2007b). Naturally, these proteins are not similar to caseins: consequently, obtained curds should be considered very different from normal raw materials, and the final product should probably show some textural defect if compared with expected cheeses. In fact, the so-called 'stretching' behaviour of cheeses is ascribed to caseins, while whey proteins have not the same property. In addition, ultrafiltration processes are particularly recommended when high moisture and unripened cheeses are desired (Banks 2007a). On the other hand, some 'fortified' curds may be obtained with the addition of low-heat treated skim milk powders (Banks 2007b).

Homogenisation of initial milks can be made for certain cheese products, but several problems—including enhanced lipolysis—may reduce its application in many situations. Consequently, these problems concern the production of initial curds. Also, high-pressure processing may be considered for safety purposes when speaking of milk for cheesemaking: in general, nutritional values and quality suffer little variations. However, similar treatments are expensive enough.

The addition of sodium chloride on fragmented curd particles is usually performed when speaking of cheddar cheeses and other 'salty' cheeses with dry salt or by immersion in brine. The dimension of curd particles may be important. Times are variable; high temperatures of brine solutions allow generally a better salt absorption (Guinee 2007; Kelly 2007b).

References

Alais C (1984) Science du lait. Principles des techniques laitières, 4th edn. S.E.P.A.I.C., Paris

Anonymous (2016) Cheese production. Milk facts, http://www.milkfacts.info/Milk%20 Processing/Cheese%20Production.htm#CStd. Accessed 05 Oct 2016

Banks JM (2007a) Why is ultrafiltration used for cheesemaking and how is it applied? In: McSweeney PLH (ed) Cheese problems solved. Woodhead Publishing Limited, Cambridge, and CRC Press LLC, Boca Raton, pp 30–33

Banks JM (2007b) How is cheese yield defined? In: McSweeney PLH (ed) Cheese problems solved. Woodhead Publishing Limited, Cambridge, and CRC Press LLC, Boca Raton, pp 102–104

Barbieri G, Barone C, Bhagat A, Caruso G, Conley ZR, Parisi S (2014) The problem of aqueous absorption in processed cheeses: a simulated approach. In: Barbieri G, Barone C, Bhagat A, Caruso G, Conley ZR, Parisi S (eds) The influence of chemistry on new foods and traditional products. Springer International Publishing, pp 1–17

Belitz HD, Grosch W, Schieberle P (2009) Food chemistry, fourth revised and extended edition, Springer, Berlin, p 529

Beresford T (2007) What problems are caused by psychrotrophs? In: McSweeney PLH (ed) Cheese problems solved. Woodhead Publishing Limited, Cambridge, and CRC Press LLC, Boca Raton, pp 13–15

Burkhalter G (1981) Catalogue of cheeses. International Dairy Federation, Brussels

Codex Alimentarius Commission (1978) Codex Stan 283-1978: general standard for cheese, rev.1-1999, amd.3-2008. Codex Alimentarius—International Food Standards, Rome

Codex Alimentarius Commission (1999) Codex Stan 208-1999: codex group standard for cheeses in Brine, amd. 2010. Codex Alimentarius—International Food Standards, Rome

Ferrari E, Gamberi M, Manzini R, Pareschi A, Persona A, Regattieri A (2003) Redesign of the Mozzarella cheese production process through development of a micro-forming and stretching extruder system. J Food Eng 59(1):13–23. doi:10.1016/S0260-8774(02)00424-7

Kelly AL (2007a) What is the typical composition of cow's milk and what milk constituents favour cheesemaking? In: McSweeney PLH (ed) Cheese problems solved. Woodhead Publishing Limited, Cambridge, and CRC Press LLC, Boca Raton, pp 3–4

Kelly AL (2007b) What potential uses do high hydrostatic pressures and high-pressure homogenisation have in cheesemaking? In: McSweeney PLH (ed) Cheese problems solved. Woodhead Publishing Limited, Cambridge, and CRC Press LLC, Boca Raton, pp 115–116

Guinee TP (2007) What are the differences between dry-salting and brine—salting? In: McSweeney PLH (ed) Cheese problems solved. Woodhead Publishing Limited, Cambridge, and CRC Press LLC, Boca Raton, pp 84–86

McSweeney PLH (2007) Principal families of cheese. In: McSweeney PLH (ed) Cheese problems solved. Woodhead Publishing Limited, Cambridge, and CRC Press LLC, Boca Raton, p 176–177

McSweeney PLH, Ottogalli G, Fox PF (2004) Diversity of cheese varieties: an overview. In: Fox PF, McSweeney PLH, Cogan TM, Guinee TP (eds) Cheese: chemistry, physics and microbiology, vol 2, 3rd edn. Elsevier Academic Press, Amsterdam, pp 1–22

O'Mahony JA, Lucey JA, McSweeney PLH (2005) Chymosin-mediated proteolysis, calcium solubilisation and texture development during the ripening of Cheddar cheese. J Dairy Sci 88:3101–3114. doi:10.3168/jds.S0022-0302(05)72992-1

Ottogalli G (2001) Atlante dei formaggi. Ulrico Hoepli Editore, Milan

Parisi S (2006) Profili chimici delle caseine presamiche alimentari. Ind Aliment 45(457):377–383

Parisi S (2016) The world of foods and beverages today. Online video course, Learning.ly/The Economist Group, 750 3rd Avenue NY. Available http://learning.ly/products/the-world-of-foods-and-beverages-today-globalization-crisis-management-and-future-perspectives

Parisi S, Barone C, Caruso G (2016) Packaging failures in frozen curds. The use of a BASIC software for mobile devices. Food Package Bull 25(1 & 2):16–19

Chapter 4
Chemical Correlations Between Industrial Curds and Final Cheeses. Can Cheesemakers Standardise Productions?

Caterina Barone, Marcella Barbera, Michele Barone, Salvatore Parisi and Izabela Steinka

Abstract The chemical composition of milks and curds influences the microbial ecology and chemical features of produced cheeses. This statement is quite obvious because 'normal' cheeses—products with a prevailing ingredient, milk—are obtained by means of the transformation of the main raw material and the addition of minor components. Because of the possible scarcity of readily available milk in many regions or economic areas, the industry of milk and dairy products, including cheeses, has improved the production of ready-to-use curds. These products, also named 'industrial curds', are produced exclusively for further cheesemaking processes. For this reason, ready-to-use curds are pre-packaged with the aim of supplying cheesemaking industries. However, the standardisation of curds is easily possible only near curd-making industries. This Chapter discusses adequate countermeasures for final cheesemakers; on the other side, the complete control on cheese production parameters in these conditions does not seem to be a possibility at present.

C. Barone (✉)
Associazione "Componiamo il Futuro" (COIF) Palermo, Palermo, Italy
e-mail: fct1970@libero.it

M. Barbera
DEMETRA Department, University of Palermo, Palermo, Italy
e-mail: marcellapu@hotmail.it

M. Barone
Associazione "Componiamo il Futuro" (COIF) Palermo, Palermo, Italy
e-mail: fct1970@libero.it

S. Parisi
Industrial Consultant, FSPCA PCQI, Palermo, Italy
e-mail: drparisi@inwind.it

I. Steinka
Gdynia Maritime University, Gdynia, Poland
e-mail: ize13ste@wp.pl

© The Author(s) 2017
C. Barone et al., *Chemical Profiles of Industrial Cow's Milk Curds*,
Chemistry of Foods, DOI 10.1007/978-3-319-50942-6_4

Keywords Cheese · Expiration date · Fat to protein ratio · Frozen storage · Industrial curd · Refrigerated storage · Standardisation

4.1   Cheeses Depend on Curds

The chemical composition and the microbial ecology of cheeses depend strongly on the original condition of used milk. This statement is quite obvious because 'normal' cheeses—products with a prevailing ingredient, milk—are obtained by means of the transformation of the main raw material and the addition of minor components. On the other hand, 'cheese-like' products include different raw materials, generally of animal origin (bovine butters, caseins, cheeses), but the 'milk' ingredient is not explicitly mentioned. In these situations, the role of original milk (if available) may be not important enough. In addition, safety problems could be solved by means of technological solutions such as melting (fusion) processes.

The production of cow's milk-made cheeses—intended in the traditional way, as 'preserved milk' (Delgado et al. 2016; Parisi 2016)—cannot be separated from the production of curds (Chap. 3). In detail, the first step of the entire production chain for the final cheese is the production of proper curds. Actually, different products require different curd with concern to moisture, fat matter on dry content, microbial counts, and other parameters. However, because of the possible scarcity of readily available milk in many regions or economic areas, the industry of milk and dairy products, including cheeses, has improved the production of different intermediates for allowed reworking purposes, such as rennet casein products, butter and creams, lactose, and ready-to-use curds, too. The last products, also named 'industrial curds', are produced exclusively for further cheesemaking processes. For this reason, ready-to-use curd is pre-packaged with the aim of supplying cheesemaking industries with limited use of transport trucks.

4.2   Cheeses and Ready-to-Use Curds. Advantages and Challenges

Cheesemakers can use this type of industrial curds because the initial raw material is quite constant when speaking of chemical composition and microbiological acceptability. On the other side, the following problems should be always considered:

(a) Ready-to-use curds have a specified 'expiration date'.
(b) Colours have to be considered carefully, because the final product depends on the main ingredient. Consequently, yellowish curds should not give white cheeses unless a minor percentage of these curds is used, and/or the final cheese is not classified as a high-moisture product. The yellow colour depends on the amount of carotenoids, generally associated with cow's milk fat molecules in

certain seasons. In fact, the higher the incorporated water, the whiter the cheese surface because white colours depend on the concentration of calcium salts, their dissolution in the 'solid aqueous solution' (the cheese), and light reflection. On the other side, white curds should not give yellow cheeses.

(c) The molecular profile of proteins (Parisi 2006). Highly demolished casein chains mean generally low water absorption, while higher molecular weights for the same proteins mean probably a better attitude to stretching techniques, higher durability, and augmented water absorption (Barbieri et al. 2014).

(d) The molecular profile of fatty chains. Lipids may be decomposed with the formation of fatty acids. Similar fatty molecules may easily be removed from the caseous matrix of curds when 'washed' with hot water. Consequently, fat matter on dry content may decrease.

(e) The 'fat to protein' ratio in curds. This ratio influences the final fat matter on dry content in produced cheeses.

(f) Possible damages on packaging and failures partially or totally correlated with packaging defects (Parisi 2013).

For these reasons, the standardisation of curds is possible near curd-making industries, while the final user—often far from the country of origin of milks, and curds—may be not able to change certain parameters. Generally, adequate countermeasures—frozen storage instead of refrigerated chillers, 'First In First Out' strategies, immediate analytical evaluations, etc.—may be helpful (Parisi et al. 2016a, b). However, the complete standardisation of cheeses in these conditions does not seem to be a possibility at present.

One option remains to be considered when speaking of a specified composition for the desired cheese and the use of a dissimilar curd (different colours or 'fat to protein' ratio, etc.). In general, average-sized cheesemaking industries are accustomed to buy different curd types, including raw materials obtained from partially skimmed milk. The last product, 'light' cow's milk curd, can be mixed with 'regular' curds (example: fat matter on dry content from 45 to 48%) with the aim of modifying the proportion between fat matter and dry content. For this reason, light curds represent a good portion of the entire production of industrial curds at present.

References

Barbieri G, Barone C, Bhagat A, Caruso G, Conley ZR, Parisi S (2014) The problem of aqueous absorption in processed cheeses: a simulated approach. In: Barbieri G, Barone C, Bhagat A, Caruso G, Conley ZR, Parisi S (eds) The influence of chemistry on new foods and traditional products, Springer International Publishing, pp 1–17

Delgado AM, Parisi S, Almeida MDV (2016) Milk and dairy products. In: Delgado AM, Almeida MDV, Parisi S (eds) Chemistry of the Mediterranean diet. Springer International Publishing, Cham, pp 139–176

Parisi S (2006) Profili chimici delle caseine presamiche alimentari. Ind Aliment 45(457):377–383

Parisi S (2013) Food Industry and packaging materials—performance-oriented guidelines for users. Smithers Rapra Technology, Shawbury

Parisi S (2016) The world of foods and beverages today. Online video course, Learning.ly/The Economist Group, 750 Third Ave NY, NY 10017. Available http://learning.ly/products/the-world-of-foods-and-beverages-today-globalization-crisis-management-and-future-perspectives

Parisi S, Barone C, Caruso G (2016a) La standardizzazione dei parametri chimici iniziali nei formaggi semiduri tramite l'uso di un programma in Basic: il CYPEP Lite, versione 1.00.1. IndAlimen 55, 568:3–6

Parisi S, Barone C, Caruso G (2016b) Packaging failures in frozen curds. The use of a BASIC software for mobile devices. Food Packag Bull 25, Numbers 1 and 2:16–19